Collins
INTERNATIONAL

T0187343

Maths
Foundation Plus
Activity Book A

Published by Collins
An imprint of HarperCollins*Publishers*
The News Building, 1 London Bridge Street,
London, SE1 9GF, UK

HarperCollins*Publishers*
Macken House, 39/40 Mayor Street Upper,
Dublin 1, DO1 C9W8, Ireland

Browse the complete Collins catalogue at
www.collins.co.uk

ISBN 978-0-00-846880-4

British Library Cataloguing-in-Publication Data
A catalogue record for this publication is available from the British Library.

Author: Peter Clarke
Publisher: Elaine Higgleton
Product manager: Letitia Luff
Commissioning editor: Rachel Houghton
Edited by: Sally Hillyer
Editorial management: Oriel Square
Cover designer: Kevin Robbins
Cover illustrations: Jouve India Pvt Ltd.
Internal illustrations: Jouve India Pvt. Ltd.
Typesetter: Jouve India Pvt. Ltd.
Production controller: Lyndsey Rogers
Printed and Bound in the UK using 100% Renewable
Electricity at Martins the Printers

Acknowledgements

With thanks to all the kindergarten staff and their schools around the world who
have helped with the development of this course, by sharing insights and
commenting on and testing sample materials:

Calcutta International School: Sharmila Majumdar, Mrs Pratima Nayar, Preeti
Roychoudhury, Tinku Yadav, Lakshmi Khanna, Mousumi Guha, Radhika Dhanuka,
Archana Tiwari, Urmita Das; Gateway College (Sri Lanka): Kousala Benedict; Hawar
International School: Kareen Barakat, Shahla Mohammed, Jennah Hussain; Manthan
International School: Shalini Reddy; Monterey Pre-Primary: Adina Oram; Prometheus
School: Aneesha Sahni, Deepa Nanda; Pragyanam School: Monika Sachdev; Rosary
Sisters High School: Samar Sabat, Sireen Freij, Hiba Mousa; Solitaire Global School:
Devi Nimmagadda; United Charter Schools (UCS): Tabassum Murtaza; Vietnam
Australia International School: Holly Simpson

The publishers wish to thank the following for permission to reproduce photographs.

(t = top, c = centre, b = bottom, r = right, l = left)

p 10tl michaeljung/Shutterstock, p 10tc StockImageFactory.com/Shutterstock,
p 10tr Andy Dean Photography/Shutterstock, p 10cl polkadot_photo/Shutterstock,
p 10c Firma V/Shutterstock, p 10cr StockImageFactory.com/Shutterstock,
p 10bl szefei/Shutterstock, p 10bc, p 10br Monkey Business Images/Shutterstock,
p 22 Shutterstock.

MIX
Paper | Supporting
responsible forestry
FSC™ C007454

This book is produced from independently certified FSC™ paper
to ensure responsible forest management.

For more information visit: www.harpercollins.co.uk/green

Trace and join

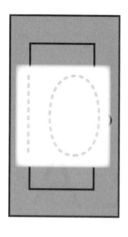

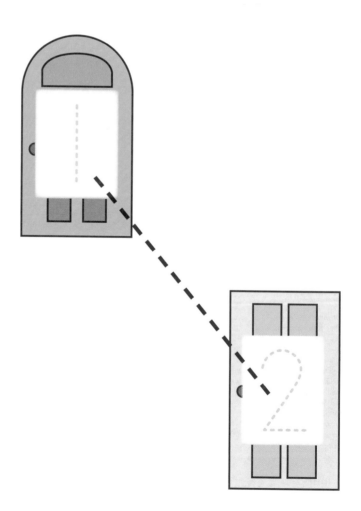

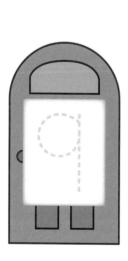

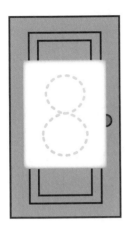

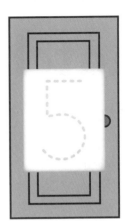

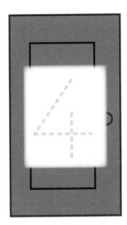

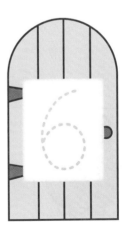

Trace the number on each door. Then draw lines
to join the doors in order, from 1 to 10.   Date:

Count and match

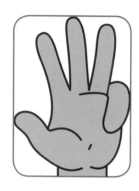

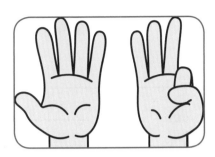

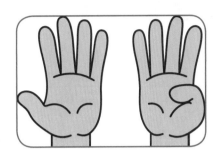

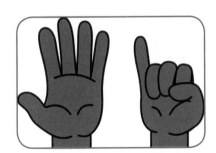

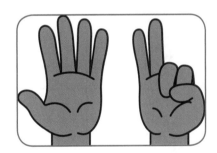

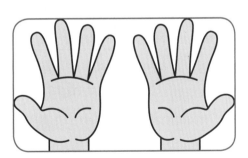

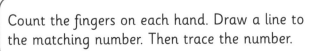

Count the fingers on each hand. Draw a line to the matching number. Then trace the number.   Date:

Fewer

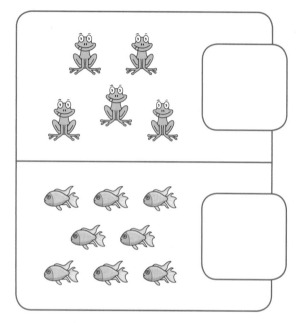

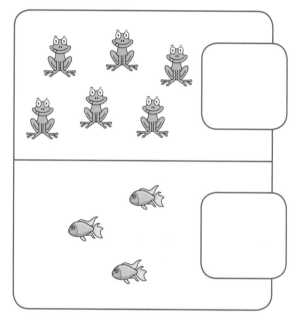

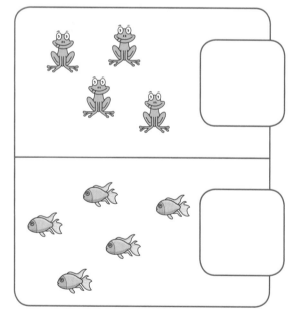

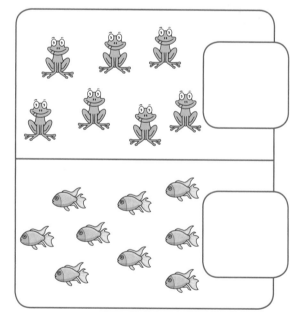

For each box, count the frogs and fish.
Put a tick next to the box that has **fewer** creatures.   Date:

More

| 1 | 2 | 3 | 4 | 5 | 6 | 7 | 8 | 9 | 10 |

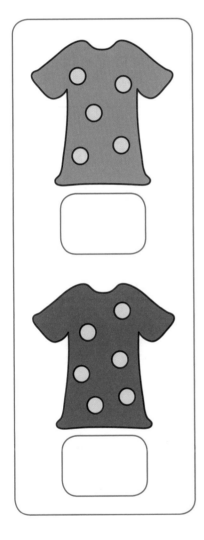
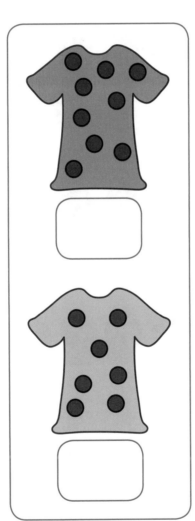
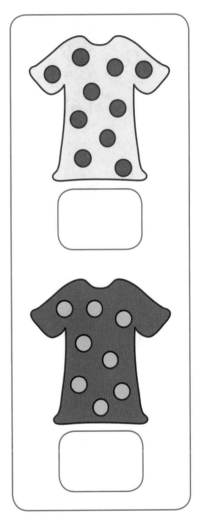

Count the spots on each T-shirt. Write the number in the box.
For each pair, circle the number that is **more**.   Date:

Ordinals

1st	2nd	3rd	4th	5th	6th	7th	8th	9th	10th

8th

1st

3rd

5th

2nd

Look at the ordinal number on the card.
Circle the child in that position.   Date:

More

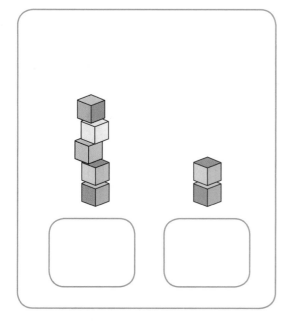

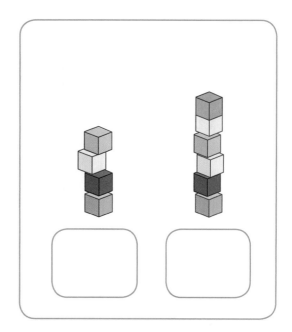

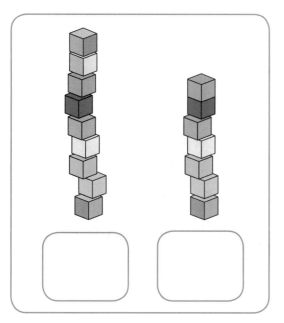

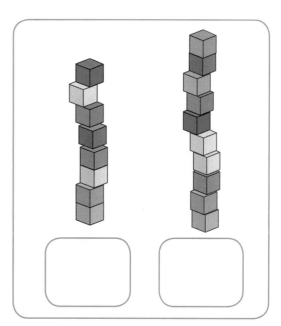

Count the blocks in each tower. Write the number in the box.
In each pair, circle the tower with **more** blocks.   Date:

Fewest and most

For each row of pictures, tick the family with the **fewest** people. Circle the family with the **most** people.

Date:

Smallest and largest

For each strip of bunting, colour the **smallest** number **red**, and the **largest** number **blue**.

Date:

Position

Colour **red** the fish in the **middle** of the tank. Colour **yellow** a fish in a **corner**. Colour **purple** a fish at an **edge**. Draw a fish **between** two fish you have coloured.   Date:

Direction

1	2	3	4	5

1	2	3	4	5

5
4
3
2
1

5
4
3
2
1

The rabbit jumps **backwards** 2 spaces. The frog jumps **forwards** 3 spaces. The bird flies **up** 2 spaces. The bee flies **down** 1 space. Circle the number each creature lands on.   Date:

Left and right

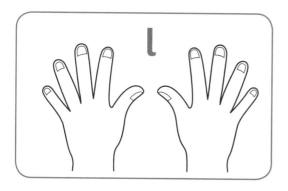

l

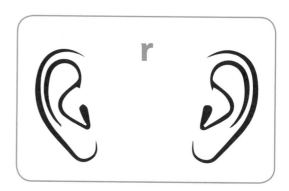

r

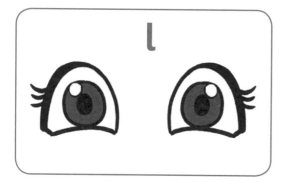

l

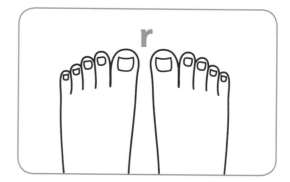

r

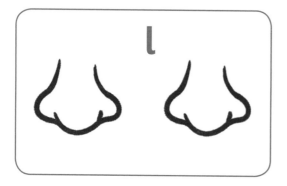

l

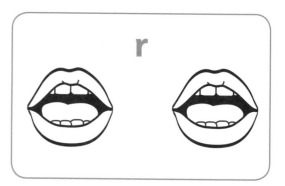

r

For the boxes on the left, colour the left-hand object. For the boxes on the right, colour the right-hand object.   Date:

Following instructions

		🐌				🪲
			🦋			
🐛	🕷		🌼	🐛		🐝
			🪰			🐜
	🦗		🐞		🦟	

left 2	up 1

down 2	right 3

Begin at the flower. For each separate instruction, circle the correct creature.

Date:

15

Longest

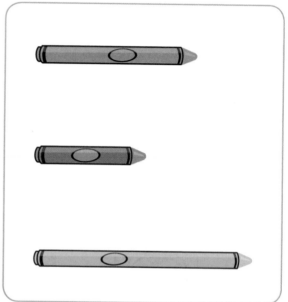

Circle the longest object or animal in each set.   Date:

Widest

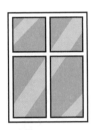  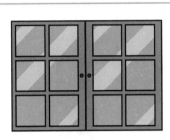

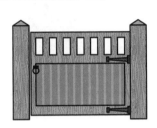 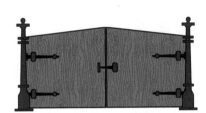 

Circle the widest object or animal in each set.   Date:

Tallest

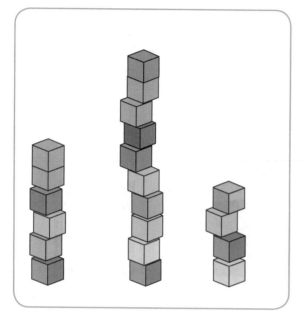

Circle the tallest object or animal in each set.   Date:

Highest

Circle the highest animal in each picture.   Date:

Sort 2D shapes

Colour **green** all the shapes with **1 side**. Colour
orange all the shapes with **3 sides**. Colour **purple**
all the shapes with **4 sides**.

Date:

Shape patterns

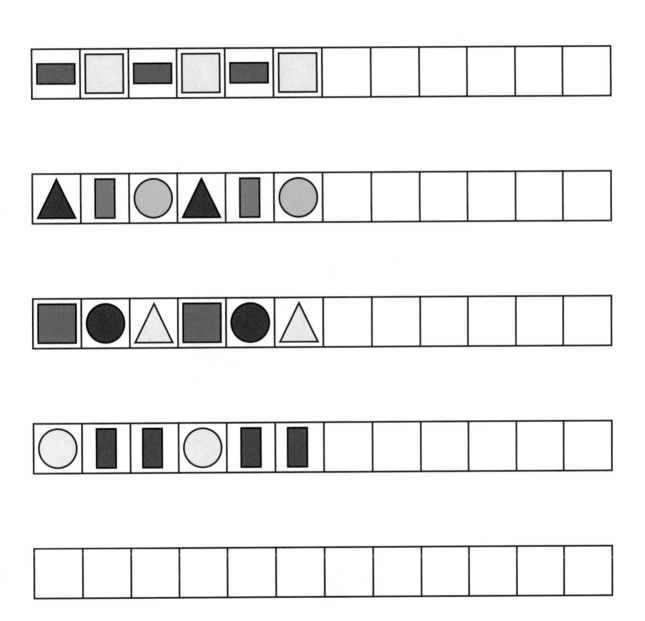

Continue each pattern. Draw your own shape pattern.   Date:

Pyramids and cones

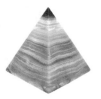

Circle all the pyramids. Count the pyramids and write the number at the top of the page. Then count the cones and write the number at the top of the page.   Date:

Sort 3D shapes

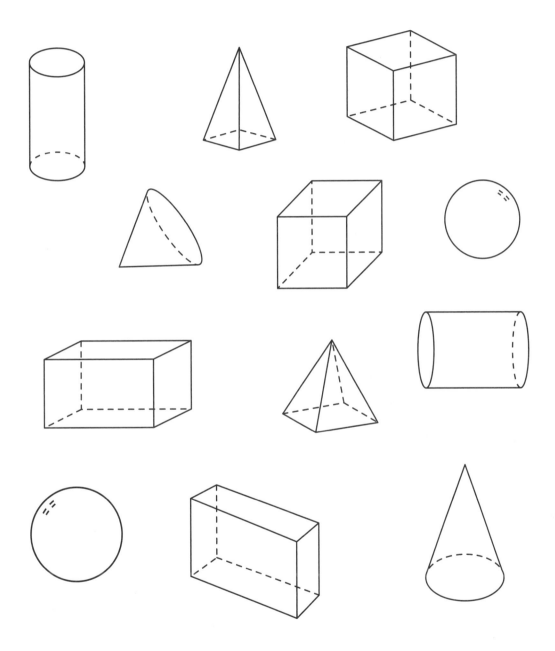

Colour **blue** all the shapes that have **6 faces**. Colour
red all the shapes that have a **curved face/surface**.   Date:

Assessment record

_____ has achieved these Maths Foundation Plus phase objectives:

Counting and understanding numbers

	1	2	3
• Count on and back in ones, starting from any number from 0 to 10.	1	2	3
• Count objects from 0 to 10.	1	2	3
• Recognise the number of objects presented in familiar patterns up to 10 without counting.	1	2	3
• Understand that zero represents none of something.	1	2	3
• Estimate a group of objects and check by counting.	1	2	3
• Understand and use ordinal numbers from 1st to 10th in different contexts.	1	2	3

Reading and writing numbers

	1	2	3
• Read and write numbers from 0 to 10.	1	2	3

Comparing and ordering numbers

	1	2	3
• Understand the relative size of quantities to compare numbers from 0 to 10.	1	2	3
• Understand the relative size of quantities to order numbers from 0 to 10.	1	2	3

Patterns and sequences

	1	2	3
• Talk about, recognise and recreate simple patterns.	1	2	3

Understanding shape

	1	2	3
• Identify, describe and sort 2D shapes, including reference to the number of sides and whether the sides are curved or straight.	1	2	3
• Identify, describe and sort 3D shapes, including reference to the number of faces and whether faces are flat or curved.	1	2	3

Position, direction and movement

	1	2	3
• Use everyday language to describe position, direction and movement.	1	2	3

Measurement

	1	2	3
• Use everyday language to describe and compare length, height and width, including long, longer, longest, short, shorter, shortest, tall, taller, tallest, wide, wider, widest, narrow, narrower and narrowest.	1	2	3

Statistics

	1	2	3
• Sort and match objects, pictures and children themselves, explaining the decisions made.	1	2	3
• Count how many objects share a particular property.	1	2	3

> 1: Partially achieved   2: Achieved   3: Exceeded

> Signed by teacher:
> Signed by parent: Date: